Josiah Keep

Common Sea-Shells of California

Josiah Keep

Common Sea-Shells of California

ISBN/EAN: 9783743338722

Manufactured in Europe, USA, Canada, Australia, Japa

Cover: Foto ©berggeist007 / pixelio.de

Manufactured and distributed by brebook publishing software
(www.brebook.com)

Josiah Keep

Common Sea-Shells of California

COMMON SEA-SHELLS

—OF—

CALIFORNIA.

—BY—

JOSIAH KEEP, A. M.

TEACHER OF NATURAL SCIENCES, ALAMEDA HIGH SCHOOL. CURATOR OF
CONCHOLOGY, CALIFORNIA ACADEMY OF SCIENCES

FULLY ILLUSTRATED BY NUMEROUS PLATES.

SAN FRANCISCO.
PRINTED FOR THE AUTHOR BY
UPTON BROS; 605 SACRAMENTO STREET.
1881.

PREFACE.

THIS little book grew out of a need, felt by the author, of a cheap and convenient manual of the more common shells found on the coast of California. Hitherto, most of the information in regard to them has been contained in rare scientific works, while good illustrations have been still more hard to obtain. The writer has therefore resolved to put the information which he has been gaining from varied sources for the last few years, in such a form that any one can use it who has the inclination.

The first chapter treats of the structure of Mollusks; the second gives directions for finding, studying, and preserving specimens; while the remainder of the book is taken up with descriptions of the several species. The work does not aim to describe all the Mollusks found on our shore, but treats especially of those common varieties which may be sought, with a reasonable hope of finding them for study and collection. Monterey Bay has been taken as the central point of study, as it furnishes a great variety of species, and its shores are becoming more and more the great summer school-ground for the students of nature.

The illustrations were drawn from nature, by the skillful pencil of Mr. Geo. H. Baker, of San Francisco, and while they constitute the most expensive part of the book, they will doubtless be found the most useful. The scientific names of species are liable to some change as the science advances, and the author begs indulgence in cases where inaccuracies occur in respect to the latest name. He has ventured, with much hesitation, to propose English names for many of the species, and trusts they may be found valuable to some, to whom the Latin names appear cumbersome.

In conclusion, the little book is dedicated to all who love to study these beautiful works of the Creator, in the hope that many more may become interested in this delightful science.

ALAMEDA, CALIFORNIA,

June, 1881.

COMMON SEA-SHELLS

—OF—

CALIFORNIA.

CHAPTER I.

SHELLS — MOLLUSKS — CLASSIFICATION — STRUCTURE OF
GASTEROPODS AND LAMELLIBRANCHS — PARTS OF THE
SHELL.

WHO has not been struck with the beauty of sea-shells? The little child has them among his choice playthings; boys and girls put them in their little cabinets; the savage uses them for money; the lady welcomes them as decorations for her parlor; the student carefully examines their structure; and the scientist reads in them the history of the growing world. Few objects in nature have been more widely diffused or more universally admired.

This general interest in shells naturally leads the thoughtful mind to study their nature and classification, and it is with a view to help those whose opportunities for such studies are few, that this little volume is prepared.

First, then, we must remember that the shell is not an inorganic body, like a stone, nor yet a mere house for an animal to live in, like the nest of a bird ; but that it is a *part* of a living creature, a hard outer layer, belonging to and produced by certain portions of the skin of the animal, somewhat as our finger nails are produced by, and really belong to the skin of the fingers. True, the creature of which the shell forms a part often has the power to withdraw its soft organs within this protection, but never, while life lasts, can it quit its shell and grow a new one.

The members of the sub-kingdom of animals to which the shell-bearers belong are called mollusks, from the Latin *mollis*, meaning "soft." Their bodies are indeed soft, since they have no internal bones to give them strength and protection ; their defense in on the outside. Some of them, it is true, have no shells. and in a few of the highest orders, the shell is internal, but none of them have true bones, nor, it may be added, true brains.

Mollusks are divided into five classes : Cephalopods, including the Nautilus and Cuttle-fish families ; Gasteropods or Creepers, such as snails and slugs ; Pteropods, little animals living mostly near the surface of the deep sea and swimming with wing-shaped paddles ; Brachiopods cr Lampshells, and Lamellibranchs. The shells of the last two classes are composed of two parts or valves, but in internal construction they are very different. The gills of the Brachiopods are arranged on long arms, while those of the Lamelli-branchs are in plates or folds, resembling in appearance a fine toothed comb. In this little book we shall speak of only the second and last classes, the other being less numerous on the sea-shore; and of less importance to the student.

The Gasteropods are more highly organized than their plate-gilled neighbors, and most of them rejoice in that chief emblem of dignity, a head. Lamellibranchs, on the other hand, being destitute of that appendage, are known as acephals, or headless animals. The acephals do not get their food by eating, but rather by breathing. Most of them live buried in the mud and communicate with the outside world by means of two pipes or siphons. Through one of these tubes they pump in water from above, then pass it over their gills, and expel it through the other siphon. The water, in passing over the long plate-gills, supplies both air and food ; first, the dissolved oxygen which it contains purifies the blood of the animal, and secondly, the great numbers of minute animal and vegetable forms contained in the water are caught by the mucous on the gills, rolled into threads, and carried to the mouth, thus furnishing the animal with its food.

The Gasteropods which live in water likewise have gills to purify their blood, but move around in search of their food, which they gather by means of a rasp-like tongue. Some of them feed on sea-weeds and other marine vegetables, while others are scavengers, clearing the sea of dead fish and the like. Not a few are predaceous, and do not scruple to attack living animals, especially the more helpless kinds of mollusks.

The Gasteropods are usually provided with elementary eyes, placed upon two tentacles or feelers; when disturbed they can quickly withdraw these eyes, and if much alarmed, all the soft parts retreat within the shell, and the aperture is closed by a little door called the *operculum*. If undisturbed for a little while, they venture out again, and crawl by means of a broad muscular disk or foot, the parts of which advance in little waves. Some mollusks construct curious egg cases, others throw their ova into the water, or hatch them in their own gills. The young generally have the power of locomotion, at least for a few hours, after which, the stationary kinds settle down for life, in their proper abodes. By means of this free early life, many species are rapidly spread over large areas. We must not omit to mention the land Gasteropods, such as snails and slugs. These breathe air by means of a simple lung, feed on vegetables, and lay their eggs under dead leaves and in similar positions.

The early naturalists classified mollusks almost wholly by their shells, but investigation has shown that some species whose shells are quite similar are very different in regard to their more vital organs, and by modern writers they have been arranged more nearly in their true position. Still, in most mollusks the shell gives pretty correct information in regard to the nature of the animal of which it forms a part. Besides, the shell is the most conspicuous and enduring part of the mollusk, and preserves its form and color indefinitely; while the soft parts must be preserved in alcohol, and then they shrivel and change their color.

For convenience sake, therefore, mollusks are described in the following pages chiefly by their shells; but the student is earnestly advised, when-ever it is possible, to carefully study the organs and habits of the living animals.

The parts of the shell may easily be learned by referring to the diagrams on Plate I. The univalve shell, Fig. 1, consists of a single tube, coiled in a spiral manner round a central axis. This tube grows larger as it advances, and usually leaves marks which indicate its stages, which are called lines of growth. The tube varies much in form in different species, some-times being flattened, sometimes angled, and again, nearly cylindrical. The opening of the tube, which often may be closed by the *opercu-lum*, is called the *aperture*, and this also varies greatly in form, being sometimes nearly round, and in other species prolonged into a tube or trough, called the *canal*. Those mollusks whose shells have this canal, (*ca, Fig.* 1), are mostly carnivorous, while most round-mouthed shells belong to vegetable feeders. Each complete turn of the spiral tube is called a *whorl*, the last and largest being the *body whorl*, and the others forming the *spire*. The point of beginning is called the *apex*, and the spiral line separating the whorls is the *suture*. The axis round which the tube revolves is called the *columella*; sometimes it is hollow for a little way, then this indentation is known as the *umbilicus*. The aperture, some-times called the mouth, is bounded on one side by the *outer lip;* the inner lip is commonly grown to the body whorl or the columella. *Spi-*

ral lines are those which follow the course of the tube; *lines of growth* are cross lines, and mark the successive positions of the outer lip. A shell is in proper position to describe when it is placed with the apex uppermost and the aperture facing the student. *Dextral* shells have the aperture on the right side; they constitute the great bulk of the univalves; a few species have the aperture on the left, and are called *sinéstral*, and occasionally a left-handed specimen of other species may be found.

In the bivalve shells, the right and left pieces, or valves, are united bp a *hinge*, which is formed of variously shaped, interlocking teeth. The office of the hinge is to hold the two valves firmly in place, and prevent their slipping in any direction. These hinge teeth are to be carefully studied, when determining the name of a species. The central teeth, found just below the *umbo*, or apex of each valve, are called *cardinal teeth;* those at the side, which are usually long and narrow, are *lateral teeth.* Near the hinge teeth is the *ligament*, or spring, composed of a dark, elastic substance, like rubber. Sometimes it is *internal*, and placed in a groove or pit. When the shell is closed, the ligament is compressed, and tends to throw the valves apart as soon as the muscles cease to act. In other shells the ligament is *external*, (see Fig. 2, Plate I), and is stretched by the closing of the two valves. In either case it tends to make the valves gape. The opposing force, which tends to close them, resides in the *adductor muscles*, which reach from one valve to the other. A few mollusks,

like the oyster, have but a single adductor, which
leaves its scar near the center of the shell. Most
of the Lamellibranchs, however, have two muscles, one near each end of the shell. These muscles, when fresh, are white and glistening, and
their strength is very considerable. They are,
in fact, the lock by which the poor mollusk fastens the door of his fortress against all intruders,
and so firm is it that it generally prevails, until
some strong-jawed fish crushes the shell, or some
insidious borer drills a hole through its hard
plates. These muscles leave markings on the
shells, which are important guides in the deter-
mination of the species.

The *mantle*, or skin of the mollusk, is partial-
ly grown to the shell, and leaves a distinct mark
called the *pallial line*, at the place where it becomes free. This line sometimes makes a deep
bend inward, called the *pallial sinus*, which is
occupied by the breathing siphons when they are
withdrawn into the shell. The part of the shell
occupied by these siphons is called the posterior,
and the other extremity the anterior portion.
When the shell is placed upon its edge, with the
posterior portion toward you, the right and left
valves correspond to your right and left hands.

In external sculpture, those lines which run
from the umbo to the edge of the valve are called *radial lines*, and those which mark the size of
the shell in its progressive stages of growth are
concentric lines. The peculiar heart-shaped depression found on some shells, beneath and in

front of the umbones, is called the *lunule*. To learn these various parts, compare the following explanation.

EXPLANATION OF PLATE I.

Fig. 1, A Univalve; *Purpura crispata. a*, apex; *sp*, spire; *ap*, aperture; *o. l*, outer lip; *ca*, canal; *u*, umbilicus; *c*, columella; *b. w*, body whorl; *s*, suture.

Fig. 2, A Bivalve, *Tapes tenerrima. u*, umbo; *lu*, lunule; *a, a*, adductor muscular impressions; *p*, pallial line; *p. s*, pallial sinus; *l. t*, lateral tooth; *c*, cardinal teeth; *l*, ligament

Mollusks, like plants, have two names; the first is the generic, or the name of the genus; the second is the specific, or the name of the species. For example, *Mytilus Californianus*, the "California mussel." By a species is meant a group of individuals which have such similarities that they may be supposed to have descended from a single pair. A genus is a group of similar species. The abbreviation after a name designates the author who first gave that name to the species.

PLATE 1.

a

Fig 1.

s

sp.

b.w.

ap.

c.

o.l.

u.

ca.

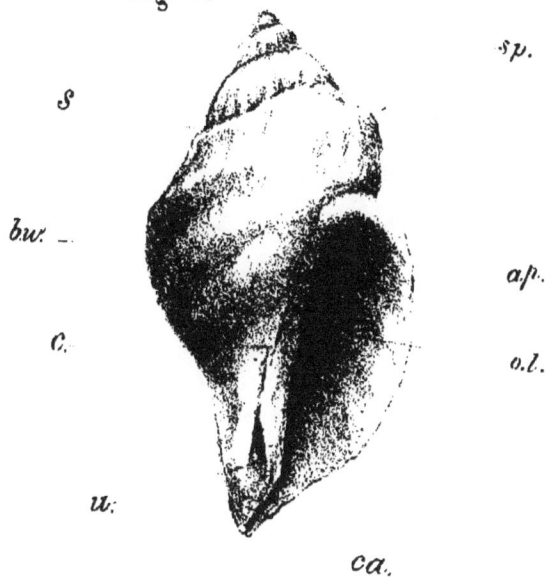

Fig. 2.

u.

l.

lu.

l.t.

c.

a'.

a.

s.

p.l.

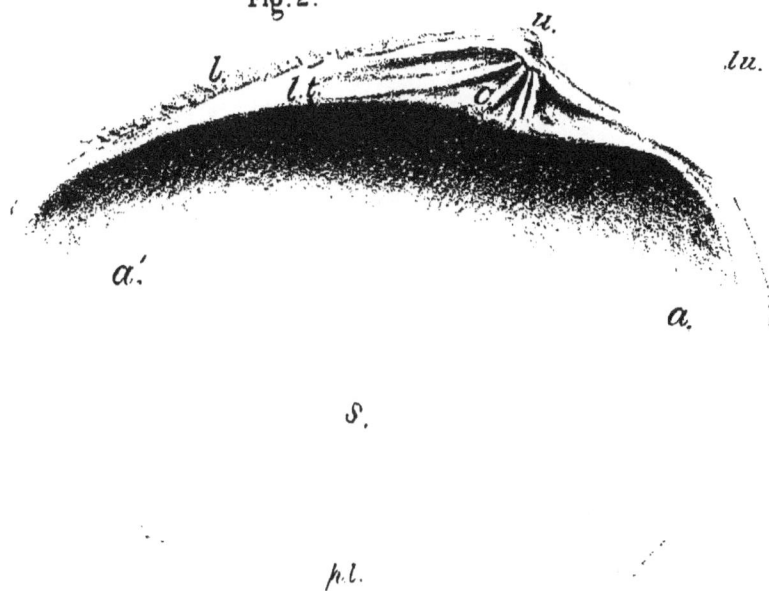

CHAPTER II.

ON COLLECTING AND PRESERVING SHELLS.

WE know but little of the creatures which live at the bottom of the deep sea, and only a few persons have an opportunity to dredge in the more shallow waters. Most of us must be content to search that narrow strip of shore which is daily left bare by the ebbing tide. But this ground will yield us many pleasant surprises if we work it carefully. To extend our field of observation as far as possible, we should take advantage of the lowest tides, such as come in the early morning. A pair of long rubber boots, and a good staff with a hook at one end are valuable assistants. *A sharp lookout should always be kept for the waves*, for many sad accidents have happened to those who have ventured too far towards the breakers, or have remained too long in dangerous clefts and caves.

The collector should follow the retreating tide as far as convenient, and carefully notice all the hiding places of the shy mollusks. Loose stones should be turned over and searched. Of course, innumerable crabs will scamper off sidewise and make an ominous rustling, but they will do no harm. Pools left by the tide often contain valuable prizes and should not be overlooked. Living mollusks are most to be sought for, but dead shells are often worth picking up. Do not mistake the brisk litle Hermit Crabs which live in empty shells for the original inhabitants; such a mistake has sometimes been made by those who have spent many years on the sea-shore.

Some bivalves, like the chamas, are found firmly attached to rocks, and so closely do they resemble the surface to which they cling that it is difficult to detect them. Others spin a byssus, or cable of threads, and anchor themselves firmly to some rock or post. Many species live in mud or sand and send up siphons to the overlying water. After the tide retreats, watch for their holes, or observe the jets of water which they sometimes throw up when disturbed. When you have found the lurking place, dig out the mollusk with a spade.

Gasteropods are usually found clinging to stones or seaweed. Some kinds can be found plenty enough, but for other species you must turn over masses of sea-moss and peer into curious cracks and crannies. The discovery of the real live animal, in his own proper home, brings a joy which is never felt by those who merely search the sandy beach, or catch their shells with the silver hook.

Limpets can easily be detached by a thin, flat knife; the more unexpected the attack, the better. Abalones must be sought in the clefts of rocks, and be dettached by a sudden pry with a long wedge. Many minute shells may be gathered by putting sea-weed into a pail of fresh

water, as the little shells are apt to fall to the
bottom. The sand of little coves should be
searched for small shells which the waves have
brought in. Some fine mollusks live on the kelp,
outside the low water mark; these may be
gathered when the sun shines, with the aid of a
boat. The stomachs of large fishes often yield
rich treasures of deep sea shells, and help us gain
some knowledge of those depths which can only
be reached by a dredge.

In collecting mollusks, carefully note their
habitat; whether they live on rocks, weeds, sand
or mud, also in what depth of water. Learn if
possible their habits and the nature of their food.
It is always well to put some of the living specimens into a pan or jar of sea water, and watch
their movements. Note carefully, in collections,
whether the shell resembles in form or color the
objects near which it is found. It has been observed by some naturalists that the imitative col-
oring of a shell apparently has much to do with its
chances of remaining unnoticed, and therefore of
preserving its existence. If this be true to a considerable extent, it may give many hints as to
the probable places in which to find certain species,

Mollusks may be preserved in alcohol entire,
but generally the shell is the only part which is
kept for the cabinet. To remove the soft parts
is an undesirable, but necessary work. It is best
done by placing the shells in boiling water for a
few minutes, and then thowing them into cold
water to cool and harden. The boiling water al-

most instantly coagulates the albumen, and loosens the connection between the mantle and the shell. The soft parts can then be removed with a little wire hook or a bent pin. If possible, they should be wholly removed, but if it cannot be done, and the shell is particularly desireable, it may be plugged with cotton. The operculum should be fastened to the cotton by a drop of mucilage. The operculum should be preserved in like manner, in all cases where a perfect shell is required. Bivalves gape after boiling, so when the flesh has been removed, the shells should be closed with a string. Chitons are hard to preserve in proper shape; it may be done, however, by tying them flat to a shingle with candle wicking, and placing them in fresh water. After they are dead, and the muscular mantle has lost its contractile power, they must be loosened from the shingle and the viscera removed with a sharp knife.

PLATE II.

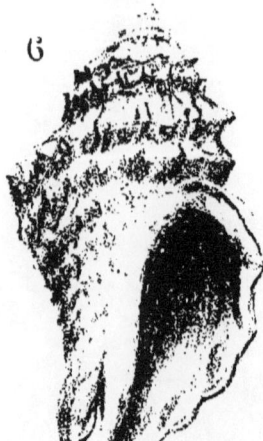

CHAPTER III.

PTERONOTUS festivus, Hds., Fig. 1, Pl. II, is a representative of the great family of the Muricidæ or Rock Shells, which abound in the warm waters around Panama, and furnish so many parlor ornaments. Most of the family have very rough exteriors, and are smooth and brilliantly tinted within. This species is about an inch and a half in length, irregularly spindle-shaped, *i. e.* tapering toward both ends, with large, reflexed frills, and a tubular canal. Color, whitish, with various dark markings. It is found chiefly in the southern part of the State.

Ranella Californica, Hds., Frog Shell, Fig. 2, Pl. II, sometimes grows to a length of six inches. It is a solid, light colored shell, with two heavy folds on opposite sides, long canal, the edge of the aperture crenulated ; spire of about four whorles. Common length, three inches.

Chrysodomus dirus, Rve., Golden Spindle-Shell, Fig. 3, Pl. II, has an ominous name, but the only ill luck we ever experienced while gathering these shells was getting caught by the tide on Duxbury Reef, at Bolinas, and being obliged to wade ashore. It is spindle shaped, from one to two

inches long, the spire consisting of five or six indistinct whorls, cut into waves by shallow furrows.
Spiral lines, numerous ; columella, enamelled ;
outer lip of a rich brown color within, with twenty light ridges. The outside of the shell is generally covered with a whitish powder, which can
be removed, showing the dark brown shell. It
may be found alive, at low tide, on the surface of
rocks which are covered with coarse sea moss.
Operculum, small and horny.

One of the most common shell on this coast is
Purpura saxicola, Val., The Rock Purple,
shown in Fig. 4, Pl. II. It derives its generic
name from the fact that the ancient Tyrians obtained a purple dye from the bodies of a similar
species, while its specific name is peculiarly ap-
propriate, for it is ever found clinging to the
rocks, and hiding in their clefts. Length, less
than an inch ; spire, short ; columella, flattened ;
outer lips, thin ; canal, short ; umbilicus, small.
The inside is reddish brown, while the outside
varies greatly both in form and color. Sometimes
it is smooth and almost black, sometimes white
and coronated, but usually it is decorated with
double spiral bands of a dark color, often accom-
panied with spiral grooves. These many varieties
probably all belong to one species. It is found
on rocks which are covered only at high tide.

Purpura canaliculata, Ducl., Grooved Purple,
Fig. 5, Pl. II, is much more rare than the last
species, probably living in deep water. In size
it is about the same, but it is more smooth and
symmetrical. The spire consists of three whorls
with a deep suture between them. From apex

to aperture run thirteen spiral grooves, giving it
the appearance of C. dirus. It can easily be distinguished by its shorter spire, smooth lip, and
deep suture. It is of light color, sometimes streak
with reddish brown.

Purpura crispata, Chem., Rough Purple, Fig.
6, Pl. II. is found in San Francisco Bay, but is
more common in more northern regions. It is
often two or three inches long, strong and heavy.
The spire consists of four strongly grooved
whorls. The body whorl is sometimes smooth,
but often very rough and foliated. It is said to
change the habit of its growth, when moved into
new localities. Its color is white or light brown.
In old shells, the umbilicus is conspicuous. Like
the other purples it has a horny, elongated operculum. Fig. 1, Pl. I, represents the same species.

Next comes a genus of mollusks, almost peculiar to the west coast of America. A good representative is shown in Fig. 1, Pl. III, *Monoceros lapilloides*, Conr., Pebbly Horn Shell. It
is so named on account of a little horn, near the
base of the outer lip. It is a pretty little shell,
found on the rocks between tides, having a spire
of four whorls, a rather small white mouth, set
with seven teeth. The outside is marked with
spiral grooves, and the lines of growth break up
the brown color into little blocks. The shells
are very thick and strong, well fitted to resist
the force of the waves. Thin and delicate shells
are generally found in deep water, where there
is little danger of being dashed against the rocks.
Shore shells are usually strong and solid. When-
ever you examine a shell, notice these points

and try to finds its adaption to its surroundings. In this way, shell-gathering becomes something more than a mere pastime, for it brings us face to face with the great questions of life, of design, and of final causes.

Monoceros engonatum, Conr., Fig. 3, Pl. III, resembles the last species in many respects, but is more angular, as its name indicates. The whorls are sharply shouldered, and the shell is smoothish and brown-dotted. By some, it is doubted whether this is more than a variety of the last species. It is more common in the southern part of the State than in the northern. A careful study of the variations of the species which are found in different localities, and which live under different conditions, is highly interesting, and accurate observations by any one may prove valuable to science. There are many things yet to be learned about our more common animals, and no one need despair of discovering new truths.

Chorus Belcheri, Hds., Fig.2, Pl. III, is a huge shell, with a long canal and a tapering spire, bristling with a crown of long, sharp points. Color, brown; length, four to six inches; more common in southern waters.

In sheltered coves the waves often wash up great numbers of little shells, which may be gathered at low tide. A very common as well as very beautiful shell which can often be thus found in great numbers is *Amphissa corrugata*, Rve., Fig. 4, Pl. III, Wrinkled Amphissa. If you search among the stones at very low tide, you will find probably find some of them alive,

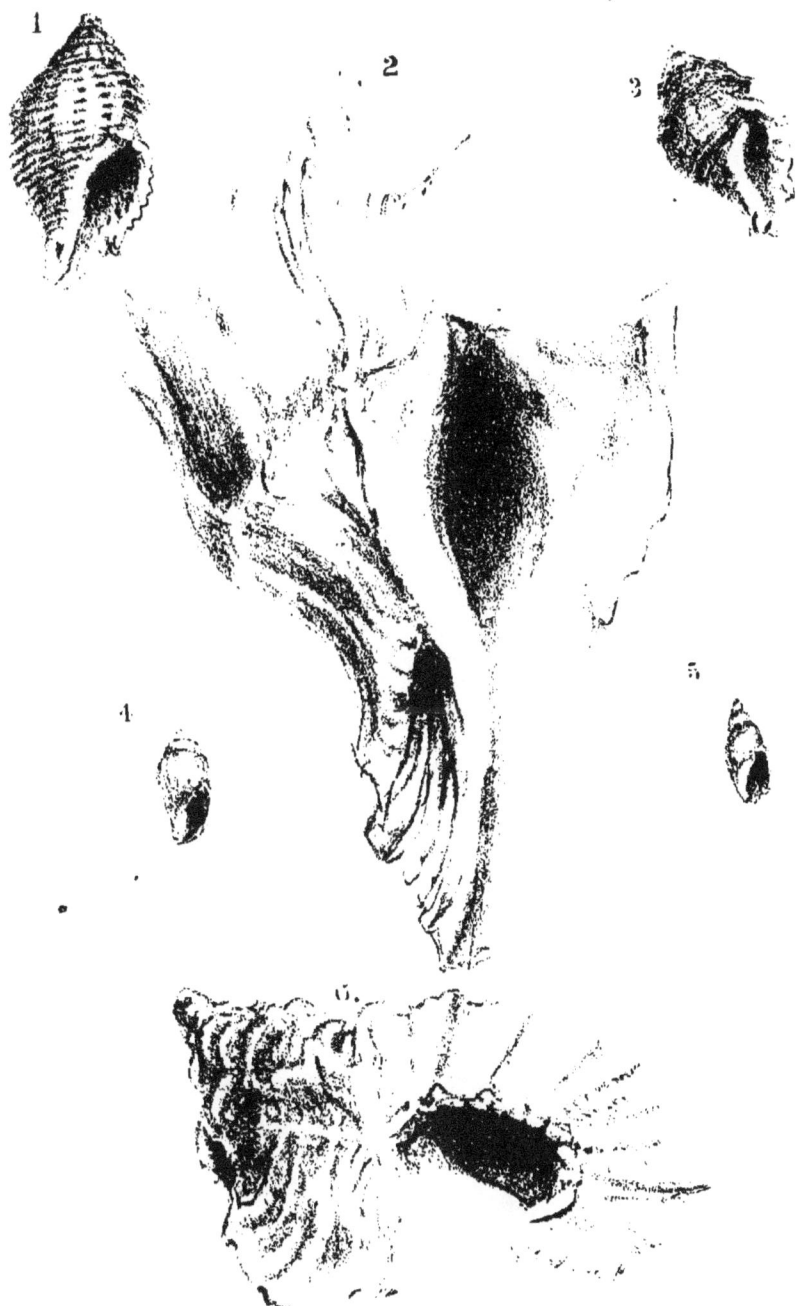

PLATE III

clinging to the rocks. Such a triumph is not soon forgotten. It is pleasant to gather dead shells on the shore, but that is not enough; you want to find the little animal at home and see how he keeps house, before you can form a correct notion of his peculiarities. Be not deceived by the little hermit crabs which love to get into dead shells and draw them around as a means of defense, but search till you find the true living mollusk. Amphissa corrugata sometimes grows to a length of one inch, but is usually about half that length. The spire consists of four whorls with a plainly marked suture. Spiral striae may be found at the base of the shell, above which the whole surface is ornamented with wavy ribs, from which it receives its name. The common color is reddish yellow, but it shades through brown to black.

Amycla carinata, Hds., Fig. 5, Pl. III, is about the size of a barley-corn. Its spire is half the length of the shell, and consists of four whorls. The body whorl has a stout keel, which gives the species its name. In some varieties this keel almost wholly disappears. Color, light brown, with a dark apex and canal; surface, smooth and glossy : may be found on the beach, with the last species.

Fig. 6, Pl. III, represents a rare and beautiful shell, *Cerostoma foliatum*, Gmel., Leafy Horn-mouth. The fine specimen which the artist used as a model was found among the rocks off Pacific Grove, at Monterey. It has three broad, winglike varices, or expansions, marking stages in growth. These varices are made up of shelly

plates which overlap like shingles. The surface
is rough and deeply sculptured. The aperture
is oval, with thin, projecting lips, which make,
by a fold, a prominent horn near its base. The
siphonal canal is long, closed, and curved at the
tip. The outside of the shell is of a dull white, and
the inside is lined with a beautiful white enamel.
The operculum looks like a thin chip of rose
wood.

Cerostoma Nuttallii, Conr. Fig. 1, Pl. IV, Nuttall's Hornmouth, resembles the last but is small-
the varices are not so broad and thin, while the
spaces between them are tubercled and marked
with but little spiral sculptuing. It was named
for the eminent naturlist, Nuttall.

The genus *Nassa*, which we will next con-
sider, has several representatives on the coast of
California. The name means "a basket for
taking fish," and refers to the netted surface of
most of the species. Among our most common
species we mention *Nassa fossata*, Gld., Basket
Shell, shown in Fig. 2, Pl. IV. It is the largest
of our species. Spire, conical, consisting of five
or six whorls ; surface, sculptured by spiral and
transverse grooves. The inside of the outer lip
is also grooved, and the aperture ends in a short,
strongly reflexed canal, through which the
animal sends up its nose-pipe, when it is search-
ing the sand for bivalves. Near the base of the
body whorl is a deep spiral ditch, or " fossa,"
which gives a name to the species. Color, yellowish white, deepening in the mouth to a brownish orange. Length, from one to two inches.

Nassa perpinguis, Hds., Fig. 3, Pl. IV, resembles the last, but is much smaller, being only three-fourths of an inch in length. Its whorls are beautifully rounded and cut into little squares. Shell, thin, light brown, with a trace of orange inside. It is found from San Francisco Bay, southward.

Nassa mendica, Gld., Fig. 4, Pl. IV, is a variable shell, about the same in length as the last, but more slender. The surface is marked with fine spiral lines and numerous transverse ribs. Its color is light brown, with a white " peristome," or margin of the aperture. Fine specimens of this species have been found at Santa Cruz and Monterey.

Nassa Cooperi, Fbs., Cooper's Basket-Shell, shown in Fig. 5, Pl. IV, is a pretty, brownish little shell, found from Bolinas Bay to San Diego. It is spirally marked, like N. mendica, but its ribs number only seven or eight to a whorl, and are quite high, giving the shell a tubercled appearance. The white lip is marked internally with small teeth.

We now come to one of the beauties of our Coast, the Purple Olive, *Olivella biplicata*, Sby. Fig. 7, Pl. IV. Every one must admire its bluish-white, polished surface, and purple mouth. The spire is short, with a distinct spiral groove separating the whorls. The inner wall of the aperture is marked by a bulge of enamel above, and two small folds beneath, which give the shell its specific name. Beds of living Olives can sometimes be found just beneath the surface of the sand, at low tide. They are active little burrowers, throwing up little ridges as they move. The shells sometimes exceed an inch in length, but commonly are smaller.

Olivella bœtica, Cpr., Slender Olive, Fig. 7, Pl. IV, has a thin, slender, brownish white shell. It is smaller and less abundant than the former species, and can easily be distinguished by its longer spire, attenuated form, and brownish color. These two species comprise all the known California Olive shells.

Conus Californicus, Hds., Fig. 1, Pl. V, is our only representative of the great Cone family, which has so many beautiful members in tropical waters. Our little species is very humble, being about an inch in length, of a chestnut color. smooth, though sometimes found covered with a hairy epidermis. The dead shells may often be found cast up on the beach.

Fig. 2, Pl. V, represents a shell often found somewhat broken, named *Drillia torosa* by Carpenter, which we may translate in part as the Knotty Drillia. Its color is from olive to black; length, from one inch to an inch and a half. spindle shaped, spire of five or six whorls, ornamented by a spiral row of knobs. It is found frequently at Monterey.

Drillia penicillata, Cpr., Fig. 3, Pl. V, Pencilled Drillia, is a very beautiful species, found at Santa Barbara and southward. It is larger and more graceful than the last named, brownish. with delicate markings, spire of eight whorls; length, one inch and a half.

Luponia (or *Cypraea*) *spadicea*, Gray, Fig. 4, Pl. V, Nut-brown Cowry, is a beautiful shell,

PLATE IV.

1

2

3.

4.

5

6.

7.

1.

2.

3.

5.

4.

6.

7.

and a worthy representative of the famous genus of Cowries. The back of the shell is well shown by the artist. It is of chestnut brown, surrounded by a darker stripe which shades off into light brown and bluish white. The other side is almost wholly occupied by the aperture and lips. The former is narrow and extends the whole length of the shell. The lips are white and set with about 22 teeth on each side. In the adult shell, the spire is completely concealed by the whorls. It was formerly quite rare, but has recently been found in considerable numbers in the southern part of the State, living with the large mussel, *Modiola modiolus*, which it greatly resembles in color. With this circumstance, Mr. R. E. C. Stearns illustrates and enforces the theory of the preserving effects of imitative color. Ordinarily, this shell would be a conspicuous object, and would quickly be taken by man or beast; when lying among the similar looking mussel shells, however, it is not easy to discover it, or distinguish it from its very different neighbors. Perhaps this influence of color has much to do with the abundance or scarcity of many other shells, and it should be carefully studied. We should ask, as we gather shells, "Do they resemble their surroundings in color or form, and does this resemblance tend to protect them?" The study of this subject may guide us to the proper place to search for shells whose color and form we know.

The little *Trivia Californica*, Gray, Fig. 5, Pl. V, is sometimes known as the Coffee-bean Shell, and its size and appearance warrant this

name. It is very plump and full, with a dozen
distinct ribs on each side.   Its color is reddish
chocolate, with white teeth on the inner wall of
the long and narrow aperture.   It is a pretty
little shell, from one-fourth to one-half inch in
length, and is sometimes worn as a jewel.

Erato vitellina, Hds., Fig. 6, Pl. V, is about
half an inch in length, quite smooth, with a
large aperture and thickened outer lip. The
spire is short ,and largely concealed.   The peristome is white, and the back is chestnut brown.
Dead shells may frequently be found along the
shore.

Fig. 7, Pl. V, represents the natural size of
the pretty little *Erato columbella,* Mke. This
species has a visible spire, long aperture with
finely toothed edges, white lips and olive back.
Still smaller, but somewhat resembling this
species, is the pure white *Marginella Jewettii,*
Cpr., not figured.   It is one-fifth of an inch long,
has a rounded spire and four distinct folds near
the base of the columella.   It resembles the
Rice Shell, *Olivella oriza,* but is shorter and
thicker.

Fig. 1, Pl. VI, represents a moderate sized
specimen of *Lunatia Lewissii,* Gld., Moon-shell.
It sometimes grows to be five or six inches in
diameter, and is a powerful enemy to helpless,
burrowing bivalves.   Plowing along through
the wet sand by means of its enormous foot, it
no sooner strikes an unfortunate clam than the
head is stretched out, and the drill which it carries in its trunk started on its errand of destruct-
ion.   Its color is yellowish white; spire, short;

form, spheroidal ; surface, smoothish ; operculum,
horny ; umbilicus, large. It somewhat resembles
the similar species, *Natica clausa*, Brod. & Sby.,
which may be distinguished by its closed umbilicus and shelly operculum.

The black, corkscrew-like shell shown in Fig.
2, Pl. VI, which we may call a Horn Shell, is
known by the name of *Cerithidea sacrata*, Gld.
There are ten whorls, numerously ribbed, with a
deep suture between them. This shell contrasts
strangely with the moon-shell, the greater part
of that being the body whorl, while there is
little to this except the spire. The outside is
dull black ; the inside, glossy brown. Length,
one, to one and a half inches. Unlike the pre-
vious species, this one delights in the brackish
waters of bays and marshes. The writer gathered them in great numbers on the muddy flats at
the head of Lake Meritt, in Oakland, where
they seemed to be enjoying the fresh air, after
the tide had left them on the surface of the mud.
Some similar species in other countries spend so
much time in the air that they have been mistaken for land shells. Owing to their form,
they are very difficult to clean. · After boiling,
as much of the soft parts as possible should be
removed, and the space plugged with cotton, to
which the horny, circular, multi-spiral operculum should be attached by a drop of glue.

Bittium filosum, Gld., Fig. 3, Pl. VI, is a
little, brownish, spiral shell, living in the sea,
and marked by spiral grooves, without ribs. It
is about half an inch long, strong and solid.

There are several other species of Bittium, most of which have ribs.

Fig. 4, Pl. VI, represents a *Scalaria* or Staircase Shell, probably of the species named *crebricostata*, "close-ribbed," by P. P. Carpenter. It is a beautiful, pure white shell, commonly smaller than the figure, marked with about 15 sharp ribs, which form a sort of crown at the suture.

Opalia borealis, Gld., Fig. 5, Pl. VI, is about an inch long, white, strong, with fewer and blunter ribs than Scalaria. It also differs from it by not having a continuous peristome.

Upon the rocks, from the region of low tide to the surface of cliffs washed by the highest spray, may be found great numbers of little, dark colored shells, about the size of peas. They belong to the genus *Littorina*, which obviously means Shore-shell. Our English cousins call them Periwinkles. The aperture to these shells is entire, and the operculum is thin, horny, and few-whorled. We have two common species, both of which are nearly black, though many specimens may be found which are spotted, striped, and even almost white.

Littorina planaxis, Nutt., Fig. 6, Pl. VI, has a short spire, round body whorl, and sharp outer lip; it is distinguished however, as its name indicates, by its flattened and scooped columella. Its length is from three-fourths of an inch downward; the interior of the mouth is brown. The other species, *Littorina scutulata*, Gld., Fig. 7, Pl. VI, is generally smaller and more pointed. The spire is as long as the aperture, the columella not excavated, and the interior of the aper-

PLATE VI.

1

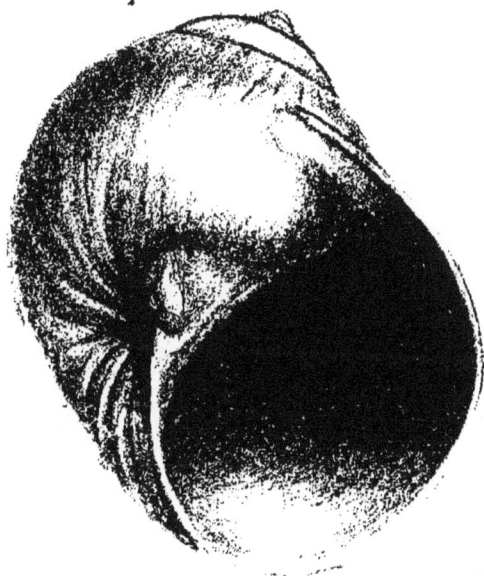

3.

4.

5

6.

7.

PLATE VII.

2.

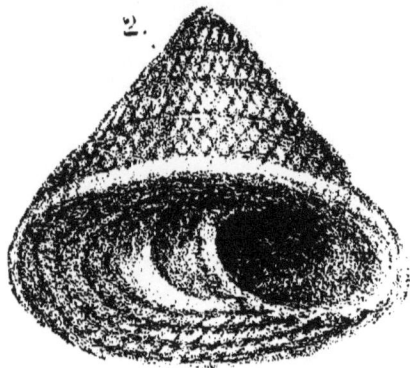

1.

3.

4.

6.

5.

ture is purple. Both of these species are very
interesting for study. They can easily be examined in little tide pools, while in motion, and
their method of clinging to the rock when the
water has left them should be noted. A near
relative to the above is the little *Lacuna unifas-
ciata*, Cpr., or Chink-shell, represented in Fig.
I, Pl. VII. It is a very little thing, about one-
sixth of an inch long, very few whorled, brown
and glossy, with the color sometimes broken into
dots on the keel of the body whorl. The aperture is semi-lunar, and the columella flattened,
with an umbilical fissure, from which it takes
its generic name. It is worth looking for.
Lacuna solidula, Lov., is sometimes half an
inch in length, but often of less size. It is
three whorled, strong, smooth, with small umbilicus, brown surface and white columella.

We have now come to the *Trochidæ* or Top-
shells, one of the most prominent and numerous
families on the coast. They vary greatly in
many respects, but still have the family traits,
and family features. Some of them are the first
to greet you as you climb down the rocks to the
shore, others are rare and shy enough; some
have little beauty, while others rival the rainbow
in their tints. They are mostly conical, with
entire apertures. and nacreous, or pearly, interi-
ors. When the outer coats are removed with
acids, the inner pearly layers appear. The animals feed on marine vegetation. We will begin
our description with the smallest species, *Phasi-
anella compta*, Gld., Pheasant-shell, shown somewhat magnified in Fig. 3, Pl. VII. It is so small

that it often escapes the collector, but when found
and examined by a glass, it shows its beauties.
Its color is white, with large zigzag stripes of
bright, cherry red.

Pachypoma gibberosum, Chem., Fig. 2, Pl.
VII, is a strong, brick-red shell, often found
dead, and commonly somewhat broken. It is
broadly conical, its whorls roughened with numerous coarse, short ribs, and its base marked by
five or six deep, concentric furrows. The operculum is oval, horny within, with a white shelly
bulge on the outside. Breadth of shell, from two
to three inches.

Fig. 4, Pl. VII, represents the large wavy
Top-shell, *Pomanlax undosus*, Wood. It sometimes grows to a size much greater than the figure, and the large animal may be seen stretched
out upon the rocks, feeding. When perfect, the
whitish pearl shell is covered with a brown, fi-
brous epidermis. In form, it is flatly conical,
with a long, triangular aperture; the outer lip is
thin, the whorls covered with undulating ribs,
and the base ornamented with beaded circles.
The operculum is very striking, horny within,
and strengthened without, by two strong, curved,
shelly ribs, as shown in the figure. This species
abounds at Santa Barbara, and southward,
though it has been found at Monterey.

In Fig. 5, Pl. VII, we can not fail to see a
model of the old friend who is so ready to greet
us as we step our feet on the rocky shore.
" Thousands of thousands " would hardly give us
a correct idea of the numbers of this Black
Turban, or *Chlorostoma funebrale*, A. Ad.

Firm and solid, well fitted to resist the buffet-
ings of the waves, it clings to the rocks which
are daily left bare. In the water, the little black
animal, with its short head and lively feelers, may
be seen crawling briskly along; while in the
air, it can wholly secrete itself within the strong
shell and close the door with its circular operculum. The color is dark purple, almost black,
with white pearly layers within; whorls, four,
often eroded at the apex; body whorl, often
puckered near the suture; umbilicus, nearly closed; columella, marked by two teeth near its base,
operculum, horny and multispiral.

Its neighbor, *Chlorostoma brunneum*, Phil.,
Brown Turban, Fig. 6, Pl. VI, is a finer species,
but is much less abundant. The best specimens
are got by hunting the kelp at low water, by
means of a boat or long rubber boots. It is of a
rich brown color, with a white mouth and very
oblique lines of growth, which give the shell a
fine striated appearance.

Chlorostoma aureotinctum, Fbs., is more flattened than the last species, with rough ribs on
the sides, and distinct grooves on the base of the
body whorl. Fig. 1, Pl. VIII, well illustrates
its form and size. Its color is dark olive, often
worn and faded; the umbilicus is conspicuous,
and touched with bright orange, which gives
rise to its pretty name. This species is generally found to the south of Monterey Bay.

The next species, *Omphalius fuscescens*, Phil.,
shown in Fig. 6, Pl. VIII, is also a southern variety. The artist has given the basal view,
showing the large umbilicus, circular aperture,

and concentric grooves. The low whorls are marked by spiral, raised lines, broken into numerous points, which give the shell a character-istic appearance; rounded teeth also line the lower part of the aperture. Its color varies from light to dark brown.

Three fine species of Top-shells, which always delight the heart of one who loves to study and collect these graceful forms of nature, have the common name Calliostoma. They are shown in Figs. 2, 3, and 4, Pl. VIII. They commonly live outside the low water mark, upon the kelp. where the rocks can not injure their delicate shells.

At times, they are very difficult to obtain, prob-ably sinking to the bottom ; but it is said that when the sun shines brightly they crawl up near the surface, and can easily be captured by the aid of a boat.

Calliostoma costatum, Mart., Fig. 2, Pl. VIII, is quite thick and strong. It has four whorls rounded and marked with fine spiral ridges. The thin, reddish brown, outer coat is easily peeled off, showing the bright, pearly blue shell underneath. The aperture of dead shells is often inhabited by the White Slipper-shell, *Crepidula navicelloides*.

Fig. 3, Pl. VIII, is a fine picture of *Calliosto-ma annulatum*, Mart., Banded Top-shell. It is a very pretty shell, thin and delicate, sharply conical, marked by fine spiral lines, which are cut into minute grains. Color, light brown, with darker dots, and a spiral line of violet. The aperture is nearly square, and the short columel-la is pearly.

PLATE VIII.

1.

2.

3.

4.

5.

7.

6.

Calliostoma canaliculatum, Mart., Grooved Top-shell, Fig. 4, Pl. VIII, is upwards of an inch in length and breadth. It is very thin and light, conical in shape, and marked by very strong spiral lines alternating with lesser ones, with a distinct suture between the five whorls. Exterior color, light brown ; interior of aperture, brightly iridescent.

Trochiscus Norrisii, Sby., Fig. 5, Pl. VIII, abundant southward, rare at Monterey, is a large, smooth, brownish shell, with low spire, rounded whorls, very large umbilicus, and ample aperture, which is closed by a mossy, circular operculum. It is found on the rocks, like the Turbans. The The last shell, No. 7, figured on plate VIII, is *Leptonyx sanguineus*, Linn., alias *Leptothyra sanguinea*. The figure shows the natural size, though it sometimes grows larger. It is strong and solid, few whorled, marked by regular spiral striæ. It may be distinguished from small specimens of C. costatum, by its rounded whorls, white interior, and white, shelly operculum. Its color is red or purple. It may be found living on the surface of rocks, at low water.

CHAPTER IV.

DESCRIPTION OF ABALONES, LIMPETS AND CHITONS

THE largest and most beautiful of our native shells next claim our attention. In the East, they are commonly called "California Shells," and are much used as decorations for mantel shelves, and bookcases. Our earliest recollection of them carries us back to the parlor of a parsonage in New England, where, after the quarterly "children's meeting," the great shell was passed around, and in it we deposited our big copper cents, which went to help establish schools for heathen children. Full well did we admire its rich, pearly lustre, and wonder at the row of mysterious holes which perforated its side. Such hallowed memories cling around the shell illustrated in Fig. 2, Pl. IX. Here in California, where they are so common, we are apt to lose some of the idea of their exceeding beauty, but in truth, they are crystallized rainbows, rich in all the tints of the spectrum.

In a commercial point of view, these shells are becoming of more and more value, vast quantities of them being worked into buttons, jewelry, inlaid and ornamental work.

Haliotis Cracherodii, Leach, commonly called
the Black Abalone, Fig. 1, Pl. IX, is the smaller
and more abundant of the two species. It may
be found from one-fourth of an inch, to six inches
in length. The back is quite smooth, marked
only by lines of growth; spire, very short; aperture, very large; holes, five to nine; color, dark
greenish black without, pearly within. It may
easily be found clinging to the rocks by its broad
foot; and when examined alive, in the water, its
black fringed mantle, stalked eyes, and slender
tentacles are most interesting for study. Note
in the shell the strong central scar, showing
where the muscle of the foot was attached to the
shell.

The Red Abalone, *Haliotis rufescens*, Swains,
Fig. 2, Pl. IX, sometimes grows to a length of
nine inches. Outer coat, red, projecting over the
inner, pearly layers, and giving the shell a red edge.
Back, somewhat roughened, often overgrown
with vegetation; holes, commonly three in number; muscular impression, prominent. Great
numbers of these shells in a partly decayed condition, may be found in the Indian shell heaps
along the coast. These may be gathered and
pulverized to form the pearl powder used in ornamenting boxes, and frames.

Haliotis splendens, Rve., is a more southern
species, found below tide, on rocks. It is more
flat, thin, and grooved, and beautifully lustrous
within. It has 4—7 holes. *Haliotis corrugata*,
Gray, also lives below tide. It is large, arched,
very rough and corrugated; holes 3—5. Fishes

eat the deep water species, and perfect shells may
sometimes be found in their stomachs.

The coast of California abounds in limpets;
they may be found in countless numbers clinging
to the rocks, their cup-shaped shells protecting
their soft bodies from injury.   They crawl along
the surface of the rock, and many of them, like
the famous birds, " come home to roost."   Some
old ones may be found settled some little dis-
tance into that spot on the rock which has long
been their home; sometimes the roost is on the
shell of another limpet, which becomes indent-
ed in the same manner.   We will first note those
which have a hole in the shell, and are thus con-
nected with the preceding genus.   *Lucapina
crenulata*, Sby., Fig. 4, Pl. IX, is a huge mollusk, six inches or more in length.   It is brick-
shaped, with a broad yellow foot, black mantle,
and has a shell on the back, largely concealed by
the folds of the mantle.   This shell, shown in
the figure, is marked by radiating ribs, crenulated edge, oblong hole, and concentric lines of
growth.   Internally, it is of a pure, glossy white,
with a thick rim of enamel around the hole.
This mollusk is not abundant.

The next Key-hole Limpet, *Fissurella volca-
no*, Rve, Fig. 3, Pl. IX, has a very suggestive
name.   It is of an oblong, conical form, smoothish, about an inch in length, with red stripes
running from the fissure to the edge, resembling
streams of hot lava from a crater.   It may occa
sionally be found living at very low water.   *Glyphis aspera*, Esch., is larger, with a small, oval
hole; white, with some color rays; very rough and

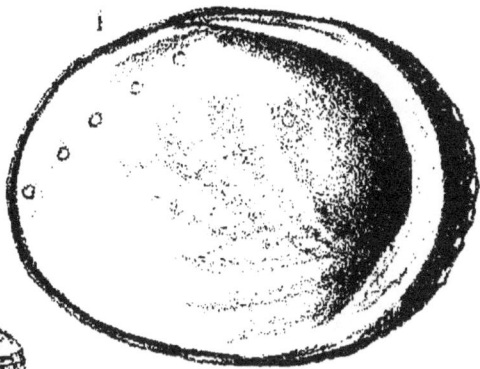

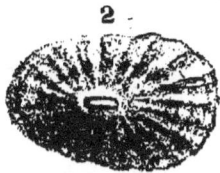

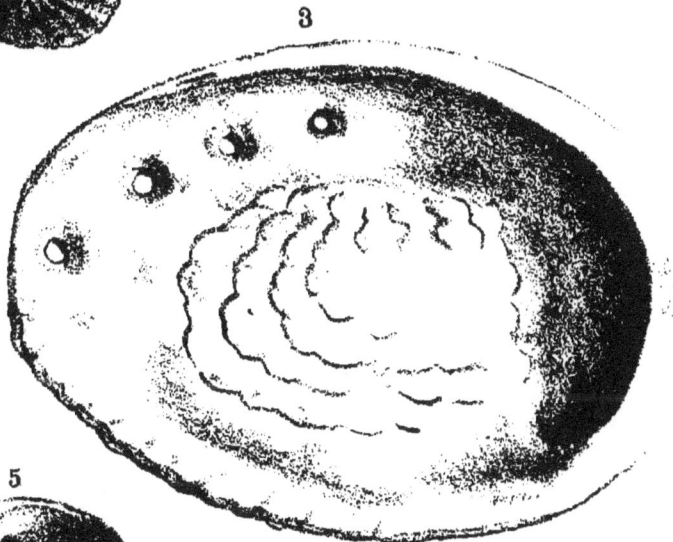

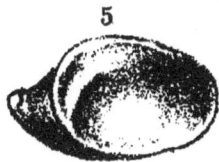

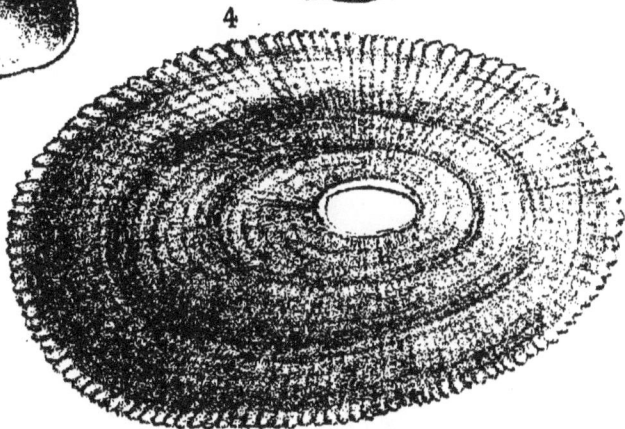

coarsely sculptured. *Glyphis densiclathrata,*
Rve., is small, white, oval, with fine, close sculpturing. *Gadinia radiata,* Cpr., has no hole. It
is circular, low conical, pure white, finely sculptured, with radiating ribs, and concentric lines
of growth. Its breadth is one-half of an inch.

The Slipper-shells are provided with a shelly
partition, partly shutting off the apex from the
aperture. We haves two very common species,
the first of which is *Crepidula adunca,* Sby.,
Hooked Slipper-shell, Fig. 5, Pl. IX. The apex
is strongly recurved; the aperture is oval and
variously distorted to fit the surface on which
the animal roosts. Its color is brown, sometimes
mottled, with a white partition inside. Its
length is from one half to a whole inch. It may
be found between tides, on rocks and shells.
The White Slipper-shell, *Crepidula navicelloides,*
Nutt., is flattened, boat-shaped, with a deck
half way across. It is very variable in shape,
adapting itself to circumstances. Small, smooth
specimens inhabit dead univalve shells, while
rough, large ones may be found on the rocks.
Color, white ; length, same as last species.

Crucibulum spinosum, Sby., "Cup and Saucer Limpet," is a more southern species. The
saucer is limpet-shaped, brownish, set with
many points or spines. Within, instead of a
deck, as in the last species, is a little triangular
cup, fastened near the apex. Size, from half an
inch to an inch and a half.

Hipponyx cranioides, Cpr., Fig. 1, Pl. X,
takes its name from its resemblance in shape to
a horse's hoof. The horse-shoe shaped muscular

scar, within the shell, is very evident. This species is thick and solid, white, more or less flattened and distorted, with rough lines of growth and indistinct radiations. Breadth, from one-half to three-quarters of an inch. The members of this genus build a shelly layer under the foot, which also shows the peculiar muscle-scar. There is a good deal of uncertainty respecting the number of our California species, as they are variable in appearance in the different stages of growth, and different species appear to run into each other. These doubtful questions present a good field for study.

Scurria mitra, Esch., Fig. 2, Pl. X, has a pure white, conical shell, which is sometimes worn as the drop of an ear-ring. It is said to live on the roots of sea-weeds. I found one living specimen on a rock, at extreme low tide. It can easily be distinguished from other shells, by its sharp conical form, pure white color, and nearly circular outline. Dead shells are common on the beach; common length, one inch.

Nacella incessa, Hds., Sea-weed Limpet, Fig. 3, Pl. X, lives on sea-weeds. It is oblong, conical, horny in appearance, brown, thin, one-half to three-fourths of an inch in length.

Most of the limpets live near the shore, between high and low water marks. Almost all of our species may be collected with moderate pains; many of them are very pretty, and they have one advantage for the collector, namely, they are easily cleaned. They may be dislodged from the rock by a sudden lift, with a broad bladed knife, but if previously disturbed, they cling

with great force, and submit to have their shells
broken before they will yield. In some countries
they are eaten, and also vast numbers are gathered as bait by the fishermen. Notice carefully
the broad foot, the mantle and gills, the short
head and tentacles, and the horse-shoe shaped
muscular impression, inside the shell.

Acmœa patina, Esch., Plate Limpet, Fig. 6,
Pl. X, is oval, flattened, with the apex nearly
central, and more or less distinct, radiating striae.
Externally it is dark, often overgrown with moss
&c; internally, near the edge, there is a dark
ring; often broken into blocks, then a broad, blue-
white enameled space, and within the muscular
impression is a patch of brown. Length of shell,
one to two inches.

Acmœa pelta, Esch , Shield Limpet, Fig. 4,
Pl. X, is more conical and pointed, with about
25 blunt ribs, sometimes obsolete. The outside
is brownish or striped; the inside white, with a
narrow, dark thread round the edge, and a brown
patch in the center. At Duxbury Reef, in Marin
Co., I found some old specimens, with very thick
shells, living in deep depressions which they had
made in the rock. A small, black, conical shell,
supposed by Carpenter to be an abnormal growth
of the young of this species, is now known as
Acmœa Asmi, Midd.

Acmœa persona, Esch., Mask Limpet, Fig. 5,
Pl. X, can easily be distinguished from the preceding species, by the posterior position of the
apex. The ribs on the front of the shell are
prominent, but rough and irregular. Its outside
color is brownish, or mottled; internally it resem-

lles A. pelta, with a wider border round the margin. This shell seldom grows more than an inch in length, and is often smaller.

Acmœa spectrum, Nutt., Fig. 7, Pl. X, is a small species, frequently found growing on the shells of other species. It is flattened, with very strong, irregular ribs, which project so as to give it a serrate edge. It is of a gray color outside, and the inner surface is irregularly marked with black and white.

Lottia gigantea, Gray., Fig. 1, Pl. IX, is a fine species, dirty brown or spotted outside, but within it is of a very dark and glossy brown, with a distinct, long, horse-shoe shaped muscular impression. The shell is long, and flattened, with the apex near one end. Length, one to three inches.

The Chitons are very strange animals, somewhat resembling limpets in shape, and like them clinging to the rocks by a broad, flat foot. Their shells, however, are not single, but composed of eight plates, which overlap like the scales on a coat of mail. These plates are surrounded and partly covered by a muscular mantle, which is often fringed and ornamented. In their early life the young chitons have little resemblance to the more adult forms, for then they consist of only two parts, head and body. Gradually the body becomes covered with seven plates, and the head takes shape and developes the eighth. Chitons are a very inoffensive class of animals, and like to live in retired places, especially under stones. Our species are quite numerous, but we will mention only those which

PLATE X.

2

3

4

5

6

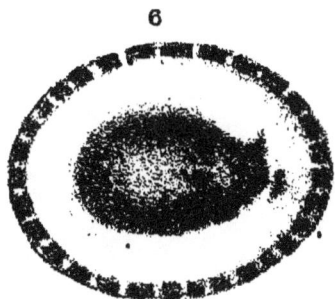

7

PLATE XI.

1

2

3

4

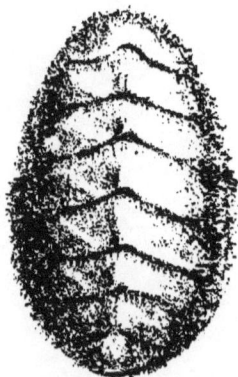

5

6

are most likely to be met with by those who
read this little book.

Ischnochiton Magdalensis, Hds., Fig. 3, Pl.
XI, Lean Chiton, is very common, and may be
found abundantly by turning over stones at low
tide. It is long and lean, as its name implies.
The valves are light colored, spotted without, and
white within. The grayish mantle is set with
minute scales. It varies in length from one to
three inches.

Mopalia muscosa, Gld., Fig. 4, Pl. XI, Mossy
Chiton, is common all along the west coast of
the United States. It is more compact than the
last species and the plates are more highly sculptured. The mantle is set with strong hairs or
points, like a chestnut bur. Outside, its color
is dark, but within the valves are of a light green.
Length, two inches.

The large chiton shown in Fig. 5, Pl. XI, was
named *Katherina Doglasiæ* by Dr. Gray of the
British Museum, in honor of Lady Katherine
Douglass, who first sent a specimen to that in-
stitution. It is now called *Katherina tunicata*,
Sby., an older specific name having been discovered and applied. If we were to put her lady-
ship's rather lengthy name into familiar English,
we could do no better, perhaps, than to call this
famous mollusk, the Black Katy. Its form is
well shown in the figure, long and oval. The
mantle is black and heavy, nearly covering the
white shell-plates ; the foot is reddish ; common
length, two to three inches. It may be found
on the rocks at the entrance to the Golden Gate,
when the tide is low, but its peculiar home is

farther northward, on the Oregon coast, and
even about Sitka. And here we may say that
a large number of the descriptions contained in
this little book apply as well to the shells of Oregon and Washington Ty., as to California. Some
of the species are much finer on the northern
coasts, than on our own. *Tonicia lineata*, Wood,
Painted Chiton. (not figured), is a small species,
about an inch in length, with smooth and naked
mantle, and eight polished valves, beautifully
striped and painted.   The chief color is orange,
and the markings are white and dark reddish
brown.   Within, the valves are white, shaded
with orange.   Living specimens are found at
very low water, but single red valves are often
washed ashore.

White, butterfly - shaped shelly plates may
often be picked up among the rocks, from one to
two inches in length ; and from their singular
appearance they have been called " Butterfly
Shells."   They are, really, single valves washed
from the dead body of the Giant Chiton, *Cryp-
tochiton Stelleri*, Midd., a huge mollusk sometimes found entire.   It is six inches in length,
and three in breadth; the white valves are
wholly concealed beneath the reddish brown,
hard, gritty mantle, which covers the whole
back of the animal.   Thus the mystery of the
" Butterfly Shells " is explained, and they become of more interest, since we know their true
origin.

The little shell shown in Fig. 2, Pl. XI, belongs to the sub-class which also includes the
Bubble-shells, and the Sea-slugs, or naked mol-

lusks. Some of the latter may be caught at low water, and they show very fine colors and painting. This species is named *Tornatella punctocœlata*, Cpr., and from its peculiar stripes is commonly called the Barrel-shell. It is about one half an inch in length, oval, very few whorled, with a fold on the columella, pure white, with two series of fine, black, spiral bands.

Bulla nebulosa, Gld., Clouded Bubble-shell, Fig. 6, Pl. XI, is a large, thin, globular shell, abounding in southern waters. Its color is brown, mottled with black and yellow clouds. The spire is depressed, leaving a hole ; length, one to two inches.

Haminea vesicula, Gld., White Bubble-shell, resembles the last in form, but is smaller, nearly white, very thin and delicate. This species lives in slimy mud, and eats animal substances.

With this shell we must take leave of the Gasteropods and turn our attention to the less highly organized, but equally interesting Lamellibranchs.

CHAPTER V.

DESCRIPTION OF BIVALVE SHELLS.

Foremost among bivalves is the oyster, the bi-valve *par excellence*. Its praises have been sung for centuries, and we fancy it will be more and m re sought for, till the end of time. Its delicate, albuminous structure, its nutritious juices, and its delicious flavor have made it a favorite dish, at least from the time of the Romans; and at the present time great numbers of men find em-ployment in propagating these mollusks, and preparing them for the table.

For study, large oysters can easily be obtained, and an examination of their organs will give a key to the anatomy of many other genera. The ligamental cavity, the single muscular impress-ion, and the structure of the hard layers can easily be studied from the dry shell, while the thin mantle, long, comb-like gills, triangular lips, and sack-shaped heart can be seen by carefully dissecting a specimen which has been killed by immersing it for some time in fresh water. An excellent little guide for such study is published by Ginn and Heath, of Boston.

Our common Oregon species, *Ostrea lurida*, Cpr., is the chief native oyster. It is small, thin,

PLATE XII.

1

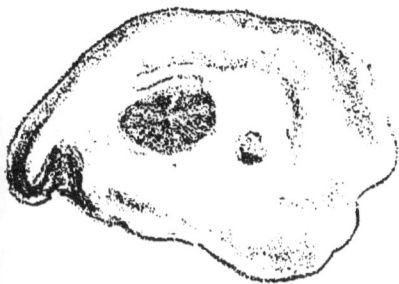

2

3

4

5

6

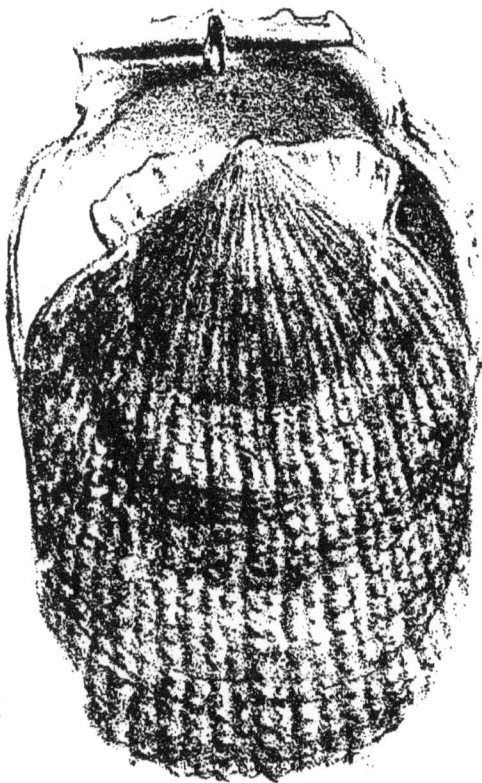

PLATE XII

.olivaceous, and sometimes stained with purple. The Eastern Oyster, *Ostrea virginiana*, is the common exotic species. They are brought here by the car load, when young, and planted in the shallow waters of San Francisco Bay. They mature in from one to three years, and though they thrive in our bays ; very few of their ova develop, probably on account of the coldness of the water.

Somewhat resembling the oyster, is the curious genus of which we have one representative, *Anomia lampe*, Gray. The shells are thin and translucent, silvery or golden. The lower valve is flat and pierced by a hole ; the upper valve is arched and marked by several muscular impressions. This species is generally found in southern waters and on the coast of Mexico, but I found one living specimen at Monterey. The shells are somewhat circular in form, about an inch and a half in diameter. A more northern species belonging to this family is *Placuanomia macroschisma*, Desh., Fig. 1, Pl. XII, gives an internal view of the upper valve of one specimen, showing the curious star-shaped muscular impression, with a smaller one below it. The form of the shell varies very greatly, though its normal shape is circular. The under valve is pierced, and both of them are strong, pearly within, and of a greenish white color. It sometimes grows to the size of a large oyster. It can always be recognized by the muscular impression, which looks like the imprint of a seal.

Of the pretty Comb-shells or Scallops, there is a good deal of variety in form and color. We venture however to illustrate two species, trust-ing that the student will carefully notice the ine-quality of the ears of the shell, and especially the little notch beneath one ear, through which the animal moors itself at times, by a series of threads, called byssus.

Pecten æquisulcatus, Cpr., Fig. 2, Pl. XII, is an inch or two in diameter, strong, full, brown-ish, and marked by about twenty equal ribs.

Pecten latiauritus, Conr., Fig. 3, Pl. XII, is very thin and delicate, having about fifteen rounded ribs. The ears are broad and unequal, the shells but little arched, while the color is brown, variously mottled with white like the feathers of a hen ; size, from one-half to one inch in diameter. These little shells live attached to sea-weed ; occasionally a storm tears up the weed and washes it ashore, thickly studded with these little beauties.

Pecten hastatus, Sby., is a much larger and stronger shell, elongated, with very unequal ears, many fine, and a few prominent ribs. Color, pinkish.

We now come to a notable California species, *Hinnites giganteus*, Gray, Fig. 4, Pl. XII, some-times called the Rock-oyster. In its early life it has a free, symmetrical shell looking like a Pecten. It is then distinguished by its very un-equal ears, and the twelve prominent, serrate ribs on the upper valve. It soon settles down for life in some convenient and sheltered spot, such as the inside of an old Haliotis shell, fastens its lower valve to this support, and yields itself up to circumstances. It soon looses its regularity of form, and becomes oyster-shaped; sometimes developing one valve and sometimes the other, as opportunity offers; twisting itself to the right or left, and becoming so distorted that it seems to have wholly forgotten its youthful grace. In color, it varies from yellow to brown, on the outside; while within, it is pure white, with a rich purple area at the hinge line. This purple color is very permanent, and fragments of shells may often be found, still showing it; the ligament is internal, in a deep, narrow pit; the muscular impression is smooth, and very large. This species sometimes grows to the size of a large oyster; it is occasionally cast up alive from deep water, but dead shells are much more common.

The last shell of this group which we shall mention is *Lima orientalis*, Ad. & Rve., Fileshell, an inside view of which is shown in Fig. 5, Pl. XII. This little shell is pure white, and is thrown up from the sea attached to sea-weed. The valves are obliquely oval, thin, gaping, and sculptured like a file; length, three-fourths of an inch.

Mytilus edulis, Linn., is the common mussel, known to every one. Spinning its strong, horny byssus, it attaches itself to rocks, posts, and wharves, in countless numbers. Its smooth, dark purple shell and orange colored soft parts are too well known to need description. The common length is two inches.

Mytilus Californianus, Conr., Fig. 6, Pl. XII, is found covering the rocks over which the breakers dash the wildest. Moored by its strong

anchor, it enjoys the rush of air and water, and
fears no danger. This species can easily be distinguished from the last by its brown, glossy
epidermis and conspicuous ribs. The shell is
purple, but more or less whitish. Some old
specimens, with nearly obsolete ribs, grow to a
length of eight or ten inches. The soft parts
are orange colored, and are frequently eaten.

Resembling the mussels are the Modiolas,
several species of which are found on the coast.
The shells are bulged near the hinge line, and
extend beyond the umbo on one side. *Modiola
capax*, Conr., is covered with a glossy, brown
epidermis, mossy in parts. The animal binds
itself by a strong byssus, and seems to seek secluded places under stones, where it can be found
at low tide. *Modiola recta*, Conr., is "long,
thin, narrow and angular, with chaff-like hairs."

Among the difficult things to explain is the
fact that a mollusk, with a thin, flexible shell,
can bore a deep hole into hard rock. That this
is done, however, can be proved by any one who
will examine the work of the Pea-pod Shell,
Adula falcata, Gld., Fig. 1, Pl. XIII. The
shell is long, narrow and slightly curved ; the inside is white and pearly, while the outside is covered with a dark chestnut epidermis, which has
numerous transverse wrinkles. I found the rocks
of Duxbury Reef, at Bolinas, almost alive with
this and other borers. The deep, narrow holes
are curved to fit the shell, and the animal also
spins a byssus, which it attaches to the sides of
its burrow ; length, two inches. *Adula stylina*,

PLATE XIII.

1

2

3

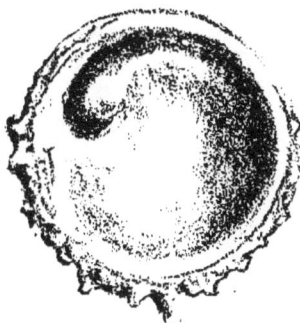

4

5

6

7

Cpr., probably a variety of this species, is shorter and broader, with a glossy epidermis. It may be found near the entrance to the Golden Gate. Another mussel-like shell, shown in Fig. 2, Pl. XIII, is *Septifer bifurcatus*, Rve. Its generic name, " partition bearer," was given from the fact that a little, internal, shelly partition stretches across each valve at the apex ; while the specific name, " two-forked," refers to the external ribs by which the shell is marked, which often divide into two. It is strong, white and nacreous within, and covered with a dark epidermis without ; the average length is one inch. It may be found, at low tide, attached to the rocks by a byssus.

The next species, *Chama pellucida*, Sby., an internal view of which is shown in Fig. 3, Pl. XIII, might easily be overlooked when it is firmly fastened to a rock. The outer surface is gray or greenish, sometimes dashed with rosy red, very rough, and marked by many close frills, which are translucent, like chalcedony. The hinge tooth is very strong, and the inside is lined with a white, opaque layer, beautifully crenulated at the edge. It grows from one to two inches in diameter, and may be found at low tide, firmly grown to the rocks, which must be broken to get the entire shell. The name " Chama," is very old, having been mentioned by Pliny.

Quite unlike the Chamas are the Cockles, or Heart-shells. Beautiful in outline, regular in growth, and free to move, they but little resemble their irregular, hap-hazard, stationary neighbors. The animal has a remarkably long foot,

with which it can dig or jump.   Their siphons
are short, so they live near the surface; they pre-
fer sandy bays, with shallow water.   *Cardium
corbis*, Mart., Fig. 4, Pl. XIII, is occasionally
found in San Francisco markets.   It is very full
and round, and varies from one to three inches in
length.   The ribs are slightly scaly, and number
about thirty.   There is a sharp cardinal tooth,
and two laterals quite distant from the umbo.
Muscular impressions, two; no pallial sinus;
edge, strongly toothed; color, whitish or light
brown.   Large broken shells are often found on
the beach near the Cliff House.

Liocardium substriatum, Conr., is a smooth
little Heart-shell from the southern part of the
State.   It is half an inch long, of a light drab
color, dotted with yellow.   Very fine lines of
growth may be observed with a glass, from which
circumstance it takes its name.   It very much
resembles a sparrow's egg.   *Liocardium elatum*,
Sby., is a very large, smooth, yellowish Cockle
from the region of San Diego.

A pure white shell, regularly marked by fine,
concentric lines is shown in Fig. 5, Pl. XIII,
and is known as *Lucina Californica*, Conr.   In
shape it is nearly circular, and varies from the
size of a dime to that of a half dollar; the cardinal teeth are small, while the lateral ones are
strong.   The ligament is external, and the small
lunule belongs wholly to the right valve.   The
anterior muscular impression is long and narrow,
and the pallial line is entire.   The pure white-
ness, symmetrical form, and regular markings
make this a very pleasing shell.   It can often be

found on the beach, thrown up by the waves.

The Kelly-shell, *Kellia Laperousii*, Desh., shown in Fig. 6, Pl. XIII, belongs to a little mollusk which shelters itself in clefts and holes, as the shell is thin and delicate. It has been mistaken for a borer, and it frequently lives in the empty holes of boring mollusks. The shell is smooth, oval, and light brown in color. It has one lateral and two cardinal hinge teeth, with a ligament between them.

A pretty little shell, not half an inch long, may sometimes be found fastened to the rocks in concealed places. An enlarged figure of it is shown in Fig. 7, Pl. XIII ; it is called *Lazaria subquadrata*, Cpr. It is strong, full, and marked by fifteen strong, rounded ribs, radiating from one angle of the four-sided shell. Its color is white, variously marked with brown spots. It is used in ornamenting shell frames.

Chione simillima, Sby., is found in the southern part of the State. Its length is two inches, and its breadth is nearly the same. The valves are very strong and thick, finely sculptured in two directions ; the radial lines are rounded, and the concentric ones are thin and sharp. Heart-shaped lunule, conspicuous; ligament, external, at the base of a broad depression ; cardinal teeth, strong, three in number ; pallial lines, entire ; color, brownish white, deeply stained inside with purple ; edge, slightly crenulated.

CHAPTER VI.

DESCRIPTION OF BIVALVE SHELLS, CONCLUDED.

Pachydesma (Trigona) *crassatelloides*, Conr., Triangle-shell, Fig. 1, Pl XIV, has a ponderous shell, sometimes growing to great size. The pair of valves from which the figure was drawn are five and one half inches in length, and weigh over a pound. They are very thick and solid, beautifully white internally, with purple muscular impressions. The outside is yellowish white, partly covered by a glossy epidermis, and sometimes marked by radiating stripes. The edges of the shell are smooth and rounded; the hinge teeth are very strong, consisting of three cardinal and one lateral ; the ligament is external and bulged ; the pallial sinus, small. This great mollusk is sometimes dug up from its home and used for food.

Another strong, thick shell, though not so large as the last, is *Saxidomus Nuttallii*, Conr., Fig. 2, Pl. XIV, Nuttall's Rock-clam. It sometimes grows to a length of four inches. The outside is of a dirty white, and is marked by numerous irregular, concentric ridges. The interior is white, and the thick part below the external ligament is translucent like agate. The

PLATE XIV.

1

3

2

5

4

pallial sinus is very deep, and the posterior end
of the shell is slightly gaping.

In all the markets of San Francisco may be
found excellent specimens of the next species,
Tapes staminea, Conr., Carpet-shell, shown in
Fig. 3, Pl. XIV. It is known among the market
men as the "Hard-shelled Clam." Tomales Bay
furnishes a good part of the supply, but it abounds
all along the coast. The valves are rounded,
full, strong, and marked by numerous narrow radiating ribs, which are cut by successive lines
of growth. It has three strongly locked cardinal hinge-teeth and an external ligament; the
pallaial sinus reaches to the middle of the shell.
The outside of some varieties is very prettily
marked by reddish brown chevrons; others are
nearly pure white. These mollusks burrow in
stony places, and can be dug out when the tide
is low. Length, one to three inches.

Fig. 5, Pl. XIV, represents one of the most
graceful of our bivalve shells, named *Amiantis
callosa*, Conr., It is pure white, full in the center
but quite thin at the edges. Its sculpturing consists of many rounded, concentric lines, equal in
size to the intervening grooves. There are no
radial markings whatever. The lunule is small,
set beneath the prominent umbones. Ligament,
external; pallial sinus, moderate; hinge, with
complicated cardinal and strong, short, lateral
teeth. This species is found to the south, and
sometimes is several inches long; common length,
two inches.

Petricola carditoides, Conr., is a Rock-shell,

as its name indicates. It would be impossible to give a figure which would illustrate all its phases, for it is exceedingly variable. Normally, it is an oval little shell, with regular sculpturing. But it has a habit of getting into a hole in the rocks and growing to fit the premises; so it frequently is much distorted, generally growing long and narrow. The ligament is external; the hinge teeth often become nearly obsolete, as well as the sculpturing, and the shell becomes thick and rough. Color, white; breadth, from one-half to three-fourths of an inch; length, one to two inches.

Tapes tenerrima, Cpr., Fig. 2, Pl. I, is often four or five inches in length. It is thin, flat, and marked by innumerable fine radiating lines, and many small concentric ridges. The cardinal hinge teeth are near the anterior extremity of the shell; the ligament is long and external; the pallial sinus is very deep. This species is sometimes thrown up from deep water, by storms. Its color is white, somewhat yellowish in places.

Rupellaria lamellifera, Conr., Fig. 4, Pl. XIV, is a nestler among rocks. It has a strong, white shell about an inch in length, and may be easily known by the ten or twelve large, thin, concentric laminæ, which mark its sides. It has no radial lines, and can thus be distinguished from the variety *ruderata*, Desh., of Tapes staminea.

The largest of all that class of mollusks commonly called clams, is known by the name *Schizothærus Nuttallii*, Conr., though by some it is called by the more simple term, "Washington Clam." It is a huge, burrowing mollusk, some-

times living two feet below the surface of the
mud. For such a situation it is provided with
an enormous siphonal tube through which it
pumps water from above. This species is used for
food, and a few of them are sufficient for an excellent chowder. The shell is oblong, bulged,
rather thin, and gapes widely where the siphons
pass. The hinge teeth are small; the ligament
is large, internal, and lodged in a triangular pit.
The pallial sinus is very broad and deep, the
lower part of it joining the ventral line. Externally the shell shows only rough lines of growth,
and is sometimes covered with an epidermis. It
delights in muddy bays, and sometimes grows
to a length of ten inches.

The Bodega Tellen, *Tellina Bodegensis*, Hinds,
Fig. 1, Pl. XV, is a very pretty sand-loving
shell, smooth, thick, and heavy, and is about two
inches long. The surface is polished, of a
creamy white color, and marked with fine concentric lines. The posterior extremity of the
shell is narrow, and somewhat bent to one side.
The ligament is external, the hinge teeth are very
small, but the pallial sinus is long and narrow.
Old specimens show a tendency to thicken the
shell from the inside.

Fig. 2, Pl. XV, represents the little shell so
often found in the sand near the Cliff House,
named *Macoma inconspicua*, Br. & Sby. It is
a thin, flat, pinkish little shell, about half an
inch long. A variety is white and larger.

Macoma secta, Conr., has a very thin, white,
glossy shell, sometimes faintly marked with red.
The ligament is short and strong placed just

back of the umbo. The hinge teeth are very
small, and the pallial sinus extends nearly the
whole length of the shell. Behind the ligament,
the shell narrows into a short, brownish wing.
The length varies from two to four inches.

Macoma nasuta, Conr., Pl. XV, Fig. 3, is a
very common species, reaching from Kamtschatka
to Mexico. It is abundant in San Francisco Bay,
and some of the large shell-mounds of the aborig-
inies, on the shores of that bay, are largely composed of the valves of this species. It inhabi's
muddy flats, burrowing quite deeply, and
reaches the water by its two, small, red siphons.
The shell is smooth and thin ; the posterior part
is narrowed and twisted to one side. The hinge
teeth are small, the ligament is external, and the
pallial sinus reaches to the forward muscular
scar. The common length is two inches.

The Red-lined Sand-shell, *Psammobia rubro-
radiata*, Nutt., is a fine shell, represented in Fig.
4, Pl. XV. The figure was drawn from a shell
found at Monterey, which measures two and one
half inches in length. The outside of the shell
is smooth and whitish, marked with broad, radiating, red bands, much resembling the rays of
the setting sun. The inside is of a fine, glossy
white, like choice porcelain. Hinge teeth, small ;
ligament, external ; sinus, large.

Sanguinolaria Nuttallii, Conr., is a southern
species. The shell is thin, rounded and flattened, covered with a glossy, dark brown epidermis,
beneath which may be seen radiating bands of
color. Ligament, large and external ; pallial
sinus, very large ; length, three inches.

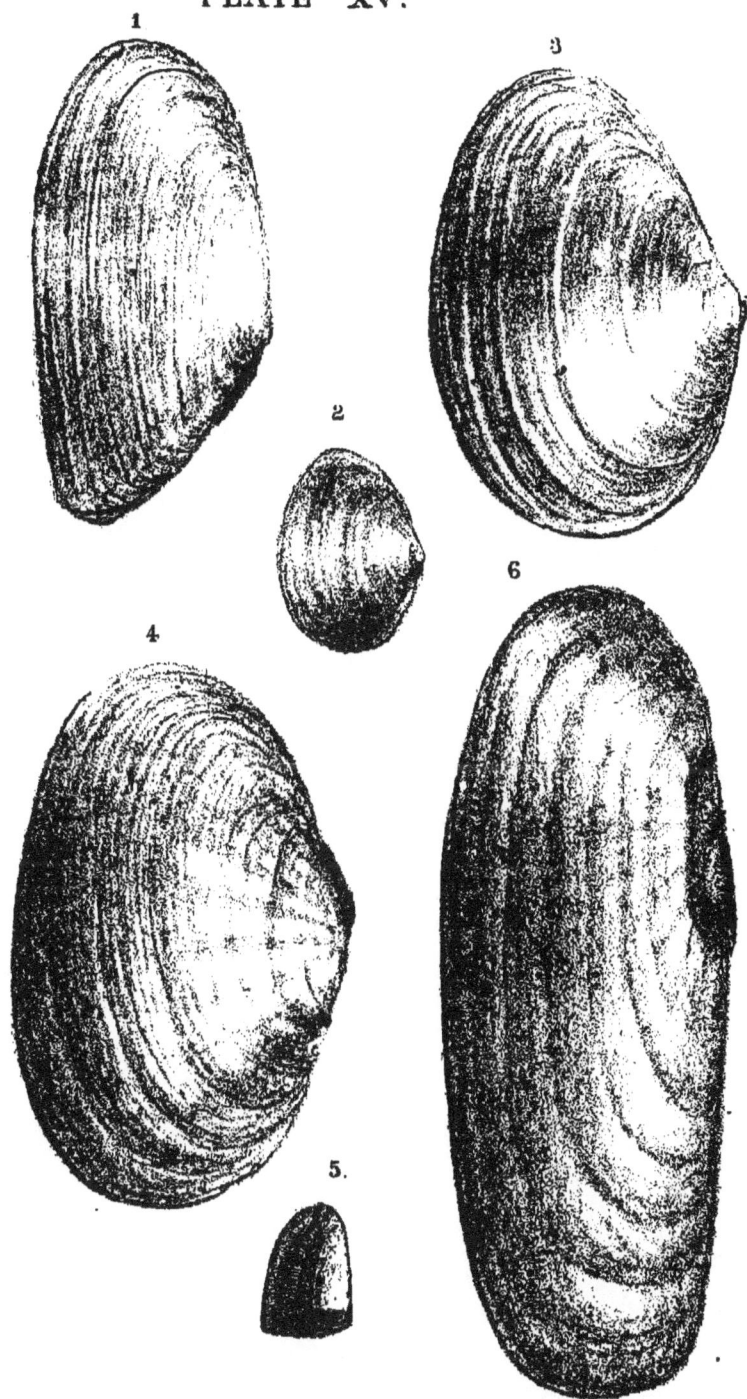

PLATE XV.

Donax Californicus, Conr., the well-known
Wedge shell, is shown in Fig. 5, Pl. XV. It is
short and stumpy, being less than an inch in
length. It varies much in color, sometimes
being nearly white, and again, striped with
bright tints. It is smooth, marked with narrow
radiations, and has a finely crenulated edge. It
abounds on the sandy beaches in the southern
part of the State.

The true Solens, or Razor-shells, are not very
numerous on the coast of California, but are
found more abundantly in the vicinity of Puget
Sound. *Solen sicarius*, Gld., is our best representative of this curious group of mollusks. The
shell of this species is about two inches long,
nearly straight, and gapes widely at both ends ;
it is thin and delicate, and is covered with a
glossy, light brown epidermis. The hinge-teeth
and ligament are very near one end of the shell.
The Solens are active burrowers, and not easily
surprised. A variety, *rosaceus*, of this species is
longer, more slender, and of a pinkish color.

Next to the true Solens, comes *Solecurtus Californianus*, Conr., or the Short Razor-shell,
shown in Fig. 1, Pl. XVI, natural size. The
epidermis of this species is not so glossy as that
of the last, and the hinge is in the middle of the
shell, instead of being at one end. The wild
ducks love to find a colony of these mollusks,
and greatly enjoy the rich feast.

Resembling the former species, but much
larger, is the beautiful shell shown in Fig. 6, Pl.
XV, known as *Machæra patula*, Dixon., Flat
Razor-shell. It grows to a length of four or five

inches, and is covered with a glossy, rich brown epidermis, which shields the thin and delicate shell. The hinge area is strengthened by a stout rib, which runs nearly across the shell. This brace presents a most evident mark of design, for it exactly fits its purpose. This species abounds in the north, and is considered delicious food. Broken shells may often be picked up near the Cliff House, in San Francisco.

Mya arenaria, Linn., well known in the markets of San Francisco, as the "Soft-shelled Clam," is not a native of this fine country, but, like the Argonauts of 1849, it came, enjoyed, settled, and multiplied. We find no trace of its shells in the old Indian mounds, and the first of the species were doubtless brought with Eastern oysters, and planted in San Francisco Bay. Unlike the oysters, however, the Mya has reproduced its kind with startling rapidity, and though it is only seven years since the first specimens were discovered in our waters, they might now be gathered by the million. Their holes may be seen all over the muddy flats, when the tide is out, and they can easily be captured by digging one or two feet deep. Although to us a solitary position in the depths of black mud would seem the acme of all disagreeable situations, our friend, M. arenaria, thrives in it remarkably, and is, no doubt, "as happy as a clam."

The shell is oblong, thin and brittle, gaping, whitish, and covered near the edge with a gray epidermis. Its most conspicuous peculiarity is the flat, spoon-shaped hinge-tooth, in the bowl of which is the ligament.

Closely resembling Mya is *Platyodon cancellatus*, Conr., Fig. 2, Pl. XVI. This genus is named from its broad hinge-tooth, which is not equal, however, to that of Mya. The hinge is near the posterior extremity of the shell, which is bulged near the umbo. Concentric markings are very plain, but radial lines are faint. It is found abundantly in Bolinas Bay. Its color is white.

Lyonsia Californica, Conr., is a delicate little shell, an inch long, which is sometimes found washed up on the shores of San Francisco Bay, as well as at other localites up and down the coast. It is oblong, bulged at one end, and pearly within. The outer coat, which shows many concentric striae, is easily rubbed off, showing the pearly layers.

We come, lastly, to the boring shells, or Piddocks. These mollusks have the power of boring holes in rocks and hardened clay. In the cabinet of the California Academy of Sciences, may be seen a piece of brick pierced by one of these borers. This shows their operations are mechanical, and not chemical. It is now pretty satisfactorily ascertained that the Piddock bores his hole by turning his shell back and forth, by means of his strong foot. The many little points and ridges cn the front of the shell act like the teeth of a file, and gradually wear away the hard rock.

First, we mention a small but common borer, *Pholadidea penita*, Conr., Fig. 3, Pl. XVI. It is seldom more than an inch and a half in length. When young, the part of the shell in front of the

file is absent, leaving the foot exposed, and in a
good condition for pressing against the side of
the burrow. In time, the shell forms a rounded
end, probably after most of the hard work of boring has been accomplished, and the animal has
a secure home. This species has curious um-
bonal reflexions, which adhere closely to the
shell. They do not work to no purpose, for
thousands of them are now engaged in filing
away dangerous reefs, along the coast.

Zirphœa crispata, Linn., Rough Piddock, Fig.
4, Pl. XVI, is a widely distributed species. It
bores in the hardest blue clay by means of its
sharp, rasp-like teeth, which are ranged in rows
on the forward part of the shell. Within the
umbo is a curious, spoon-shaped plate or tooth.
A little supplemental plate covers the hinge area.
The shell is white, thin and brittle, and it gapes
widely at both ends ; its common length is two
or three inches.

The last species which we will mention is the
Great California Piddock, *Parapholas Californica*, Conr., shown in Fig. 5, Pl. XVI. This
noble species is some three or four inches in
length. The anterior portion is rounded and
full, while the posterior parts taper to small di-
mensions, and end in brown flaps of epidermis,
which project beyond the shell. The rocky
dust which the animal excavates is utilized in
building up a strong, thick, conical chimney,
which protects the siphons

THE END

PLATE XVI.

1

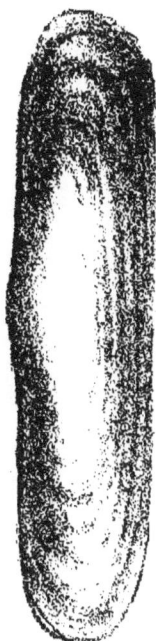

2

3

4

5

INDEX TO ILLUSTRATIONS.

INDEX.

ERRATA.

Page 6, line 19; for " in," read is.   Page 8, line 6 ; for "o-erclum," read op.rculum.   Page 24, line 32; for " Cypraca," read Cypraea.